© Houghton Mifflin Harcourt Publishing Company • Cover Image Credits: (Bighorn Sheep)©Blaine Harrington III/ Alamy Images; (Watchman Peak, Utah) ©Russ Bishop/Alamy Images

Made in the United States
Text printed on 100%
recycled paper

Houghton Mifflin Harcourt

GO MATH!

Printed in the U.S.A.

ISBN 978-0-544-34246-0

15 16 17   0928   22 21 20 19 18

4500742674   C D E F G

Dear Students and Families,

Welcome to **Go Math!**, Grade 6! In this exciting mathematics program, there are hands-on activities to do and real-world problems to solve. Best of all, you will write your ideas and answers right in your book. In **Go Math!**, writing and drawing on the pages helps you think deeply about what you are learning, and you will really understand math!

By the way, all of the pages in your **Go Math!** book are made using recycled paper. We wanted you to know that you can Go Green with **Go Math!**

Sincerely,

The Authors

Made in the United States
Text printed on 100% recycled paper

GO MATH!

Authors

Juli K. Dixon, Ph.D.
Professor, Mathematics Education
University of Central Florida
Orlando, Florida

Edward B. Burger, Ph.D.
President, Southwestern University
Georgetown, Texas

Steven J. Leinwand
Principal Research Analyst
American Institutes for
 Research (AIR)
Washington, D.C.

Contributor

Rena Petrello
Professor, Mathematics
Moorpark College
Moorpark, California

Matthew R. Larson, Ph.D.
K-12 Curriculum Specialist for
 Mathematics
Lincoln Public Schools
Lincoln, Nebraska

Martha E. Sandoval-Martinez
Math Instructor
El Camino College
Torrance, California

English Language Learners Consultant

Elizabeth Jiménez
CEO, GEMAS Consulting
Professional Expert on English
 Learner Education
Bilingual Education and
 Dual Language
Pomona, California

Ratios and Rates

Critical Area Connecting ratio and rate to whole number multiplication and division and using concepts of ratio and rate to solve problems

6 Units of Measure 313

COMMON CORE STATE STANDARDS

6.RP Ratios and Proportional Relationships
Cluster A Understand ratio concepts and use ratio reasoning to solve problems.
6.RP.A.3d

GO DIGITAL

Go online! Your math lessons are interactive. Use *i*Tools, Animated Math Models, the Multimedia eGlossary, and more.

Chapter 6 Overview

In this chapter, you will explore and discover answers to the following **Essential Questions**:

• How can you use measurements to help you describe and compare objects?

• Why do you need to convert between units of measure?

• How can you use a ratio to convert units?

• How do you transform units to solve problems?

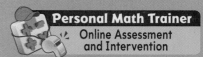

Personal Math Trainer
Online Assessment and Intervention

CRITICAL AREA REVIEW PROJECT THE GREAT RACE: *www.thinkcentral.com*

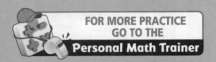
FOR MORE PRACTICE
GO TO THE
Personal Math Trainer

Practice and Homework

Lesson Check and
Spiral Review in
every lesson

Units of Measure

© Houghton Mifflin Harcourt Publishing Company • Image Credits: (b) ©DLILLC/Corbis

✔ Show What You Know

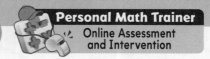
Personal Math Trainer
Online Assessment
and Intervention

Check your understanding of important skills.

Name _____

▶ **Choose the Appropriate Unit** Circle the more reasonable unit to measure the object. (4.MD.A.1)

1. the length of a car
   inches or feet

2. the length of a soccer field
   meters or kilometers

▶ **Multiply and Divide by 10, 100, and 1,000** Use mental math. (5.NBT.A.2)

3. 2.51×10

_____

4. 5.3×100

_____

5. $0.71 \times 1,000$

_____

6. $3.25 \div 10$

_____

7. $8.65 \div 100$

_____

8. $56.2 \div 1,000$

_____

▶ **Convert Units** Complete. (5.MD.A.1)

9. $12 \text{ lb} = \blacksquare \text{ oz}$
   Think: 1 lb = 16 oz

_____

10. $8 \text{ c} = \blacksquare \text{ pt}$
   Think: 2 c = 1 pt

_____

11. $84 \text{ in.} = \blacksquare \text{ ft}$
   Think: 12 in. = 1 ft

_____

A cheetah can run at a rate of 105,600 yards per hour. Find the number of miles the cheetah could run at this rate in 5 minutes.

Vocabulary Builder

Sort the review words into the Venn diagram.

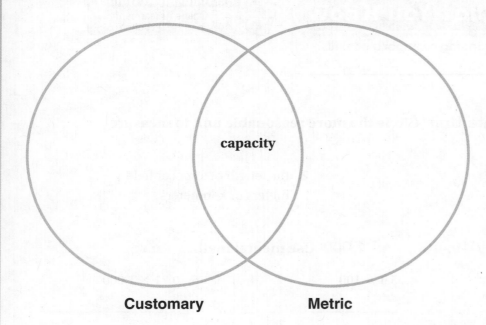

Customary   Metric

Review Words

✓ gallon
  gram
✓ length
  liter
✓ mass
  meter
  ounce
  pint
  pound
✓ quart
  ton
✓ weight

Preview Words

✓ capacity
✓ conversion factor

▶ **Understand Vocabulary** •

Complete the sentences by using the checked words.

1. A rate in which the two quantities are equal but use different

    units is called a _____.

2. _____ is the the amount of matter in an object.

3. _____ is the amount a container can hold.

4. The _____ of an object tells how heavy the
    object is.

5. Inches, feet, and yards are all customary units used to measure

    _____.

6. A _____ is a larger unit of capacity than a quart.

GO DIGITAL
• **Interactive Student Edition**
• **Multimedia eGlossary**

Chapter 6 Vocabulary

capacity

capacidad

9

common factor

factor común

12

conversion factor

factor de conversión

16

denominator

denominador

20

formula

fórmula

34

numerator

numerador

66

square unit

unidad cuadrada

96

weight

peso

107

A number that is a factor of two or more numbers

Example:

Factors of 16: 1, 2, 4, 8, 16

Factors of 20: 1, 2, 4, 5, 10, 20

The amount a container can hold

Examples: $\frac{1}{2}$ gallon, 2 quarts

The number below the bar in a fraction that tells how many equal parts are in the whole or in the group

$$\frac{3}{4} \longleftarrow \text{denominator}$$

A rate in which two quantities are equal, but use different units

Example: 1 yard = 3 feet, so use the rate $\frac{1 \text{ yd}}{3 \text{ ft}}$ to convert yards from feet

---

The number above the bar in a fraction that tells how many equal parts of the whole are being considered

$$\frac{3}{4} \longleftarrow \text{numerator}$$

A set of symbols that expresses a mathematical rule

Example: Use $d = r \times t$ to find distance.

How heavy an object is

Example:

soccer ball
14–16 ounces

A unit used to measure area such as square foot (ft²), square meter (m²), and so on

Bingo

For 3–6 players

Materials

- 1 set of word cards
- 1 Bingo board for each player
- game markers

How to Play

1. The caller chooses a card and reads the definition. Then the caller puts the card in a second pile.
2. Players put a marker on the word that matches the definition each time they find it on their Bingo boards.
3. Repeat Steps 1 and 2 until a player marks 5 boxes in a line going down, across, or on a slant and calls "Bingo."
4. To check the answers, the player who said "Bingo" reads the words aloud while the caller checks the definitions.

Word Box

capacity

common factor

conversion factor

denominator

formula

numerator

square unit

weight

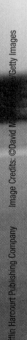

The Write Way

Reflect

Choose one idea. Write about it.

- Describe a situation in which you might use a conversion factor.
- Sumeer has a jar that holds 4 cups of water. Eliza has a bottle that holds 3 pints of water. Explain whose container has the greater capacity.
- If the area of a room is 20 feet by 6 yards, tell how to figure out the room's area in square feet.
- Write and solve a word problem that uses the formula $d = r \times t$.

Convert Units of Length

Essential Question How can you use ratio reasoning to convert from one unit of length to another?

Common Core **Ratios and Proportional Relationships—6.RP.A.3d**
MATHEMATICAL PRACTICES
MP2, MP6, MP8

In the customary measurement system, some of the common units of length are inches, feet, yards, and miles. You can multiply by an appropriate conversion factor to convert between units. A **conversion factor** is a rate in which the two quantities are equal, but use different units.

Customary Units of Length

1 foot (ft) = 12 inches (in.)
1 yard (yd) = 36 inches
1 yard = 3 feet
1 mile (mi) = 5,280 feet
1 mile = 1,760 yards

 Unlock the Problem Real World

In a soccer game, Kyle scored a goal. Kyle was 33 feet from the goal. How many yards from the goal was he?

Math Idea

When the same unit appears in a numerator and a denominator, you can divide by the common unit before multiplying as you would with a common factor.

 Convert 33 feet to yards.

Choose a conversion factor. **Think:** I'm converting *to* yards *from* feet.

1 yard = 3 feet, so use the rate $\frac{1 \text{ yd}}{3 \text{ ft}}$.

Multiply 33 feet by the conversion factor. Units of *feet* appear in a numerator and a denominator, so you can divide the units before multiplying.

$$33 \text{ ft} \times \frac{1 \text{ yd}}{3 \text{ ft}} = \frac{33 \text{ ft}}{1} \times \frac{1 \text{ yd}}{3 \text{ ft}} = \underline{\hspace{1cm}} \text{ yd}$$

So, Kyle was _____ yards from the goal.

 How many inches from the goal was Kyle?

Choose a conversion factor. **Think:** I'm converting *to* inches *from* feet.

12 inches = 1 foot, so use the rate $\frac{12 \text{ in.}}{1 \text{ ft}}$.

Multiply 33 ft by the conversion factor.

$$33 \text{ ft} \times \frac{12 \text{ in.}}{1 \text{ ft}} = \frac{33 \text{ ft}}{1} \times \frac{12 \text{ in.}}{1 \text{ ft}} = \underline{\hspace{2cm}} \text{ in.}$$

So, Kyle was _____ inches from the goal.

 Math Talk

MATHEMATICAL PRACTICES ⑥

Explain How do you know which unit to use in the numerator and which unit to use in the denominator of a conversion factor?

Metric Units You can use a similar process to convert metric units. Metric units are used throughout most of the world. One advantage of using the metric system is that the units are related by powers of 10.

Metric Units of Length

1,000 millimeters (mm)	= 1 meter (m)
100 centimeters (cm)	= 1 meter
10 decimeters (dm)	= 1 meter
1 dekameter (dam)	= 10 meters
1 hectometer (hm)	= 100 meters
1 kilometer (km)	= 1,000 meters

Example A passenger airplane is 73.9 meters long. What is the length of the airplane in centimeters? What is the length in kilometers?

One Way Use a conversion factor.

73.9 meters = ■ centimeters

Choose a conversion factor.

100 cm = 1 m, so use the rate $\dfrac{\blacksquare \; cm}{\blacksquare \; m}$.

Multiply 73.9 meters by the conversion factor. Simplify the common units before multiplying.

$$\frac{73.9 \; \cancel{m}}{1} \times \frac{\blacksquare \; cm}{\blacksquare \; \cancel{m}} = \underline{\hspace{2cm}} \; cm$$

So, 73.9 meters is equal to _____ centimeters.

ERROR Alert

Be sure to use the correct conversion factor. The units you are converting from should simplify to 1, leaving only the units you are converting to.

Another Way Use powers of 10.

Metric units are related to each other by factors of 10.

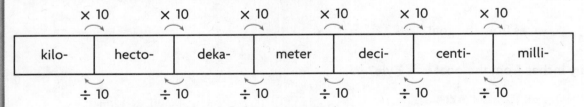

kilo-	hecto-	deka-	meter	deci-	centi-	milli-

73.9 meters = ■ kilometers

Use the chart.

Kilometers are 3 places to the left of meters in the chart. Move the decimal point 3 places to the left. This is the same as dividing by 1,000.

73.9   0.0739

So, 73.9 meters is equal to _____ kilometer.

Math Talk

MATHEMATICAL PRACTICES ❷

Reasoning If you convert 285 centimeters to decimeters, will the number of decimeters be greater or less than the number of centimeters? Explain.

316

© Houghton Mifflin Harcourt Publishing Company

Name _____

Convert to the given unit.

1. 3 miles = ▪ yards

   conversion factor: $\dfrac{\quad\quad\text{yd}}{\text{mi}}$

   3 miles = $\dfrac{3\text{ mi}}{1} \times \dfrac{1{,}760\text{ yd}}{1\text{ mi}}$ = _____ yd

2. 43 dm = _____ hm

✓3. 9 yd = _____ in.

4. 72 ft = _____ yd

✓5. 7,500 mm = _____ dm

MATHEMATICAL PRACTICES ⑧

Generalize How do you convert from inches to yards and yards to inches?

On Your Own

6. Rohan used 9 yards of ribbon to wrap gifts. How many inches of ribbon did he use?

_____

7. One species of frog can grow to a maximum length of 12.4 millimeters. What is the maximum length of this frog species in centimeters?

_____

8. The height of the Empire State Building measured to the top of the lightning rod is approximately 443.1 meters. What is this height in hectometers?

_____

9. **GO DEEPER** A snail moves at a speed of 2.5 feet per minute. How many yards will the snail have moved in half of an hour?

_____

Practice: Copy and Solve Compare. Write <, >, or =.

10. 32 feet ◯ 11 yards

11. 537 cm ◯ 5.37 m

12. 75 inches ◯ 6 feet

Problem Solving · Applications Real World

What's the Error?

13. **THINK SMARTER**   The Redwood National Park
is home to some of the largest trees in the world.
Hyperion is the tallest tree in the park, with a
height of approximately 379 feet. Tom wants
to find the height of the tree in yards.

Tom converted the height this way:

$$3 \text{ feet} = 1 \text{ yard}$$

conversion factor: $\frac{3 \text{ ft}}{1 \text{ yd}}$

$$\frac{379 \text{ ft}}{1} \times \frac{3 \text{ ft}}{1 \text{ yd}} = 1,137 \text{ yd}$$

Find and describe Tom's error.	**Show how to correctly convert from 379 feet to yards.**

So, 379 feet = _____ yards.

- **MATHEMATICAL PRACTICE 6** **Explain** how you knew Tom's answer was incorrect.

_____

_____

14. **THINK SMARTER**   Choose $<$, $>$, or $=$.

14a.   12 yards   [$<$ / $>$ / $=$]   432 inches   14b.   321 cm   [$<$ / $>$ / $=$]   32.1 m

Convert Units of Length

COMMON CORE STANDARD—6.RP.A.3d
Understand ratio concepts and use ratio reasoning to solve problems.

Convert to the given unit.

1. 42 ft = [ ] yd

conversion factor: $\dfrac{1 \text{ yd}}{3 \text{ ft}}$

42 ft × $\dfrac{1 \text{ yd}}{3 \text{ ft}}$

42 ft = 14 yd

2. 2,350 m = [ ] km

_____

3. 18 ft = [ ] in.

_____

4. 289 m = [ ] dm

_____

5. 5 mi = [ ] yd

_____

6. 35 mm = [ ] cm

_____

Compare. Write <, >, or =.

7. 1.9 dm ◯ 1,900 mm

8. 12 ft ◯ 4 yd

9. 56 cm ◯ 56,000 km

10. 98 in. ◯ 8 ft

11. 64 cm ◯ 630 mm

12. 2 mi ◯ 10,560 ft

Problem Solving · Real World

13. The giant swallowtail is the largest butterfly in the United States. Its wingspan can be as large as 16 centimeters. What is the maximum wingspan in millimeters?

_____

14. The 102nd floor of the Sears Tower in Chicago is the highest occupied floor. It is 1,431 feet above the ground. How many yards above the ground is the 102nd floor?

_____

15. **WRITE** ▸*Math* Explain why units can be simplified first when measurements are multiplied.

_____

_____

Lesson Check (6.RP.A.3d)

1. Justin rides his bicycle 2.5 kilometers to school. Luke walks 1,950 meters to school. How much farther does Justin ride to school than Luke walks to school?

2. The length of a room is $10\frac{1}{2}$ feet. What is the length of the room in inches?

Spiral Review (6.NS.C.8, 6.RP.A.3a, 6.RP.A.3c)

3. Each unit on the map represents 1 mile. What is the distance between the campground and the waterfall?

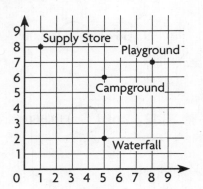

4. On a field trip, 2 vans can carry 32 students. How many students can go on a field trip when there are 6 vans?

5. According to a 2008 survey, $\frac{29}{50}$ of all teens have sent at least one text message in their lives. What percent of teens have sent a text message?

6. Of the students in Ms. Danver's class, 6 walk to school. This represents 30% of her students. How many students are in Ms. Danver's class?

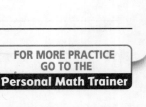

FOR MORE PRACTICE
GO TO THE
Personal Math Trainer

Convert Units of Capacity

Essential Question How can you use ratio reasoning to convert from one unit of capacity to another?

 Common Core **Ratios and Proportional Relationships—6.RP.A.3d**
MATHEMATICAL PRACTICES
MP2, MP6

Capacity measures the amount a container can hold when filled. In the customary measurement system, some common units of capacity are fluid ounces, cups, pints, quarts, and gallons. You can convert between units by multiplying the given units by an appropriate conversion factor.

Customary Units of Capacity

8 fluid ounces (fl oz)	= 1 cup (c)
2 cups	= 1 pint (pt)
2 pints	= 1 quart (qt)
4 cups	= 1 quart
4 quarts	= 1 gallon (gal)

 Unlock the Problem Real World

A dairy cow produces about 25 quarts of milk each day. How many gallons of milk does the cow produce each day?

- How are quarts and gallons related?

  _____

- Why can you multiply a quantity by $\frac{1\ gal}{4\ qt}$ without changing the value of the quantity?

  _____

  _____

🔑 **Convert 25 quarts to gallons.**

Choose a conversion factor. **Think:** I'm converting *to* gallons *from* quarts.

1 gallon = 4 quarts, so use the rate $\frac{1\ gal}{4\ qt}$.

Multiply 25 qt by the conversion factor.

$$25\ qt \times \frac{1\ gal}{4\ qt} = \frac{25\ qt}{1} \times \frac{1\ gal}{4\ qt} = 6\ \frac{}{4}\ gal$$

The fractional part of the answer can be renamed using the smaller unit.

$$6\ \frac{}{4}\ gal = \underline{\quad}\ gallons,\ \underline{\quad}\ quart$$

So, the cow produces _____ gallons, _____ quart of milk each day.

🔑 **How many pints of milk does a cow produce each day?**

Choose a conversion factor. **Think:** I'm converting *to* pints *from* quarts.

2 pints = 1 quart, so use the rate $\frac{\ pt}{\ qt}$.

Multiply 25 qt by the conversion factor.

$$25\ qt \times \frac{\ pt}{\ qt} = \frac{25\ qt}{1} \times \frac{\ pt}{\ qt} = \underline{\quad}\ pt$$

So, the cow produces _____ pints of milk each day.

Metric Units You can use a similar process to convert metric units of capacity. Just like metric units of length, metric units of capacity are related by powers of 10.

Metric Units of Capacity

1,000 milliliters (mL) = 1 liter (L)
100 centiliters (cL) = 1 liter
10 deciliters (dL) = 1 liter
1 dekaliter (daL) = 10 liters
1 hectoliter (hL) = 100 liters
1 kiloliter (kL) = 1,000 liters

🔑 Example A piece of Native American pottery has a capacity of 1.7 liters. What is the capacity of the pot in dekaliters? What is the capacity in milliliters?

🔑 One Way Use a conversion factor.

1.7 liters = ▇ dekaliters

Choose a conversion factor.

1 dekaliter = 10 liters, so use the rate

$$\frac{\boxed{}\ \text{daL}}{\boxed{}\ \text{L}}.$$

Multiply 1.7 L by the conversion factor.

$$\frac{1.7\ \cancel{L}}{1} \times \frac{\boxed{}\ \text{daL}}{\boxed{}\ \cancel{L}} = \underline{\hspace{2cm}}\ daL$$

So, 1.7 liters is equivalent to _____ dekaliter.

🔑 Another Way Use powers of 10.

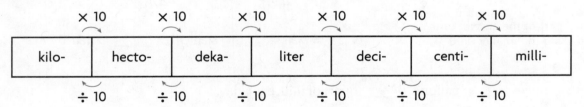

	× 10	× 10	× 10	× 10	× 10	× 10
kilo-	hecto-	deka-	liter	deci-	centi-	milli-
	÷ 10	÷ 10	÷ 10	÷ 10	÷ 10	÷ 10

1.7 liters = ▇ milliliters

Use the chart.

Milliliters are 3 places to the right of liters. So, move the decimal point 3 places to the right.

1.7 1700.

So, 1.7 liters is equal to _____ milliliters.

Math Talk

MATHEMATICAL PRACTICES ❻

Explain Why can't you convert between units in the customary system by moving the decimal point left or right?

• **MATHEMATICAL PRACTICE ❻ Describe a Method** Describe how you would convert kiloliters to milliliters.

Name _____

Share and Show

Convert to the given unit.

1. 5 quarts = ▇ cups

    conversion factor: $\dfrac{\text{c}}{\text{qt}}$

    5 quarts = $\dfrac{5\,\text{qt}}{1} \times \dfrac{4\,\text{c}}{1\,\text{qt}}$ = _____ c

2. 6.7 liters = _____ hectoliters

✓ 3. 5.3 kL = _____ L

✓ 4. 36 qt = _____ gal

5. 5,000 mL = _____ cL

MATHEMATICAL PRACTICES 6

Compare the customary and metric systems. In which system is it easier to convert from one unit to another?

On Your Own

6. It takes 41 gallons of water for a washing machine to wash a load of laundry. How many quarts of water does it take to wash one load?

    _____

7. Sam squeezed 237 milliliters of juice from 4 oranges. How many liters of juice did Sam squeeze?

    _____

8. **MATHEMATICAL PRACTICE 2 Reason Quantitatively** A bottle contains 3.78 liters of water. Without calculating, determine whether there are more or less than 3.78 deciliters of water in the bottle. Explain your reasoning.

    _____

    _____

9. **GO DEEPER** Tonya has a 1-quart, a 2-quart, and a 3-quart bowl. A recipe asks for 16 ounces of milk. If Tonya is going to triple the recipe, what is the smallest bowl that will hold the milk?

    _____

    _____

Practice: Copy and Solve Compare. Write <, >, or =.

10. 700,000 L ◯ 70 kL

11. 6 gal ◯ 30 qt

12. 54 kL ◯ 540,000 dL

13. 10 pt ◯ 5 qt

14. 500 mL ◯ 50 L

15. 14 c ◯ 4 qt

Unlock the Problem

16. **THINK SMARTER** Jeffrey is loading cases of bottled water onto a freight elevator. There are 24 one-pint bottles in each case. The maximum weight that the elevator can carry is 1,000 pounds. If 1 gallon of water weighs 8.35 pounds, what is the maximum number of full cases Jeffrey can load onto the elevator?

a. What do you need to find?

_____

_____

_____

b. How can you find the weight of 1 case of bottled water? What is the weight?

_____

_____

_____

_____

_____

c. How can you find the number of cases that Jeffrey can load onto the elevator?

_____

_____

_____

_____

_____

_____

_____

_____

_____

d. What is the maximum number of full cases Jeffrey can load onto the elevator?

_____

_____

17. **GO DEEPER** Monica put 1 liter, 1 deciliter, 1 centiliter, and 1 milliliter of water into a bowl. How many milliliters of water did she put in the bowl?

_____

18. **THINK SMARTER** Select the conversions that are equivalent to 235 liters. Mark all that apply.

(A) 235,000 milliliters

(B) 0.235 milliliters

(C) 235,000 kiloliters

(D) 0.235 kiloliters

Convert Units of Capacity

Common Core **COMMON CORE STANDARD—6.RP.A.3d**
Understand ratio concepts and use ratio reasoning to solve problems.

Convert to the given unit.

1. 7 gallons = [ ] quarts

conversion factor: $\frac{4\ qt}{1\ gal}$

7 gal × $\frac{4\ qt}{1\ gal}$

7 gal = 28 qt

2. 5.1 liters = [ ] kiloliters

Move the decimal point **3** places to the left.

5.1 liters = **0.0051** kiloliters

3. 20 qt = [ ] gal

4. 40 L = [ ] mL

5. 33 pt = [ ] qt [ ] pt

6. 29 cL = [ ] daL

7. 7.7 kL = [ ] cL

8. 24 fl oz = [ ] pt [ ] c

Problem Solving Real World

9. A bottle contains 3.5 liters of water. A second bottle contains 3,750 milliliters of water. How many more milliliters are in the larger bottle than in the smaller bottle?

10. Arnie's car used 100 cups of gasoline during a drive. He paid $3.12 per gallon for gas. How much did the gas cost?

11. **WRITE** *Math* Explain how units of length and capacity are similar in the metric system.

_____

_____

Lesson Check (6.RP.A.3d)

1. Gina filled a tub with 25 quarts of water. What is this amount in gallons and quarts?

2. Four horses are pulling a wagon. Each horse drinks 45,000 milliliters of water each day. How many liters of water will the horses drink in 5 days?

Spiral Review (6.NS.C.8, 6.RP.A.2, 6.RP.A.3b, 6.RP.A.3c, 6.RP.A.3d)

3. The map shows Henry's town. Each unit represents 1 kilometer. After school, Henry walks to the library. How far does he walk?

4. An elevator travels 117 feet in 6.5 seconds. What is the elevator's speed as a unit rate?

5. Julie's MP3 player contains 860 songs. If 20% of the songs are rap songs and 15% of the songs are R&B songs, how many of the songs are other types of songs?

6. How many kilometers are equivalent to 3,570 meters?

FOR MORE PRACTICE
GO TO THE
Personal Math Trainer

Convert Units of Weight and Mass

Essential Question How can you use ratio reasoning to convert from one unit of weight or mass to another?

Ratios and Proportional Relationships—6.RP.A.3d

MATHEMATICAL PRACTICES
MP1, MP2, MP3, MP4

The weight of an object is a measure of how heavy it is. Units of weight in the customary measurement system include ounces, pounds, and tons.

Customary Units of Weight
1 pound (lb) = 16 ounces (oz)
1 ton (T) = 2,000 pounds

🔑 Unlock the Problem

The largest pearl ever found weighed 226 ounces. What was the pearl's weight in pounds?

- How are ounces and pounds related?

  _____

- Will you expect the number of pounds to be greater than 226 or less than 226? Explain.

  _____

 Convert 226 ounces to pounds.

Choose a conversion factor.
Think: I'm converting *to* pounds *from* ounces.

1 lb = 16 oz, so use the rate $\dfrac{\boxed{}\ \text{lb}}{\boxed{}\ \text{oz}}$.

Multiply 226 ounces by the conversion factor.

$$226 \text{ oz} \times \frac{1 \text{ lb}}{16 \text{ oz}} = \frac{226 \cancel{\text{oz}}}{1} \times \frac{1 \text{ lb}}{16 \cancel{\text{oz}}} = \frac{\boxed{}}{16} \text{ lb}$$

Think: The fractional part of the answer can be renamed using the smaller unit.

$$\frac{\boxed{}}{16} \text{ lb} = \underline{\quad} \text{ lb}, \underline{\quad} \text{ oz}$$

So, the largest pearl weighed _____ pounds, _____ ounces.

 The largest emerald ever found weighed 38 pounds. What was its weight in ounces?

Choose a conversion factor.
Think: I'm converting *to* ounces *from* pounds.

16 oz = 1 lb, so use the rate $\dfrac{\boxed{}\ \text{oz}}{\boxed{}\ \text{lb}}$.

Multiply 38 lb by the conversion factor.

$$38 \text{ lb} \times \frac{16 \text{ oz}}{1 \text{ lb}} = \frac{38 \cancel{\text{lb}}}{1} \times \frac{16 \text{ oz}}{1 \cancel{\text{lb}}} = \underline{\qquad\qquad} \text{ oz}$$

So, the emerald weighed _____ ounces.

1. **MATHEMATICAL PRACTICE ④ Model Mathematics** Explain how you could convert the emerald's weight to tons.

_____

Metric Units The amount of matter in an object is called the mass. Metric units of mass are related by powers of 10.

Metric Units of Mass
1,000 milligrams (mg) = 1 gram (g)
100 centigrams (cg) = 1 gram
10 decigrams (dg) = 1 gram
1 dekagram (dag) = 10 grams
1 hectogram (hg) = 100 grams
1 kilogram (kg) = 1,000 grams

Example
Corinne caught a trout with a mass of 2,570 grams. What was the mass of the trout in centigrams? What was the mass in kilograms?

One Way Use a conversion factor.

2,570 grams to centigrams

Choose a conversion factor.

100 cg = 1 g, so use the rate $\dfrac{\quad cg}{\quad g}$.

Multiply 2,570 g by the conversion factor.

$$\dfrac{2,570\ \cancel{g}}{1} \times \dfrac{100\ cg}{1\ \cancel{g}} = \underline{\hspace{2cm}}\ cg$$

So, the trout's mass was _____ centigrams.

Another Way Use powers of 10.

Recall that metric units are related to each other by factors of 10.

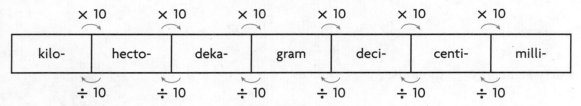

×10	×10	×10	×10	×10	×10
kilo-	hecto-	deka-	gram	deci-	centi-	milli-
÷10	÷10	÷10	÷10	÷10	÷10

2,570 grams to kilograms

Use the chart.

Kilograms are 3 places to the left of grams. Move the decimal point 3 places to the left.

2570.   2.570

So, 2,570 grams = _____ kilograms.

 Math Talk

Reason Quantitatively Compare objects with masses of 1 dg and 1 dag. Which has a greater mass? Explain.

2. **MATHEMATICAL PRACTICE ①** **Describe Relationships** Suppose hoots and goots are units of weight, and 2 hoots = 4 goots. Which is heavier, a hoot or a goot? Explain.

_____

_____

Name _____

Convert to the given unit.

1. 9 pounds = [ ] ounces

   conversion factor: $\dfrac{oz}{lb}$

   9 pounds = $9 \text{ lb} \times \dfrac{16 \text{ oz}}{1 \text{ lb}}$ = _____ oz

2. 3.77 grams = _____ dekagram

3. Amanda's computer weighs 56 ounces. How many pounds does it weigh?

   _____

4. A honeybee can carry 40 mg of nectar. How many grams of nectar can a honeybee carry?

   _____

Math Talk MATHEMATICAL PRACTICES ③

Compare How are metric units of capacity and mass alike? How are they different?

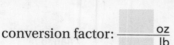 **On Your Own**

Convert to the given unit.

5. 4 lb = _____ oz

6. 7.13 g = _____ cg

7. 3 T = _____ lb

8. The African Goliath frog can weigh up to 7 pounds. How many ounces can the Goliath frog weigh?

   _____

9. The mass of a standard hockey puck must be at least 156 grams. What is the minimum mass of 8 hockey pucks in kilograms?

   _____

Practice: Copy and Solve Compare. Write <, >, or =.

10. 250 lb ◯ 0.25 T

11. 65.3 hg ◯ 653 dag

12. 5 T ◯ 5,000 lb

13. THINK SMARTER Masses of precious stones are measured in carats, where 1 carat = 200 milligrams. What is the mass of a 50-dg diamond in carats?

Problem Solving • Applications Real World

Use the table for 14–17.

14. Express the weight range for bowling balls in pounds.

_____

15. GO DEEPER How many more pounds does the heaviest soccer ball weigh than the heaviest baseball? Round your answer to the nearest hundredth.

_____

16. THINK SMARTER A manufacturer produces 3 tons of baseballs per day and packs them in cartons of 24 baseballs each. If all of the balls are the minimum allowable weight, how many cartons of balls does the company produce each day?

_____

17. MATHEMATICAL PRACTICE 5 **Communicate** Explain how you could use mental math to estimate the number of soccer balls it would take to produce a total weight of 1 ton.

_____

_____

_____

_____

18. THINK SMARTER The Wilson family's newborn baby weighs 84 ounces. Choose the numbers to show the baby's weight in pounds and ounces.

5		3	
6	pounds	4	ounces
7		5	

Sport Ball Weights (in ounces)	
baseball 5–5.25	handball 2.1–2.3
bowling ball 160–256	soccer ball 14–16

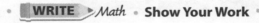WRITE Math • **Show Your Work**

Name _____

Convert Units of Weight and Mass

COMMON CORE STANDARD—6.RP.A.3d
Understand ratio concepts and use ratio reasoning to solve problems.

Convert to the given unit.

1. 5 pounds = ⬚ ounces

conversion factor: $\frac{16\ oz}{1\ lb}$

5 pounds = 5 $\cancel{lb}$ × $\frac{16\ oz}{1\ \cancel{lb}}$ = **80** oz

2. 2.36 grams = ⬚ hectograms

Move the decimal point **2** places to the left.

2.36 grams = **0.0236** hectogram

3. 30 g = ⬚ dg

4. 17.2 hg = ⬚ g

5. 400 lb = ⬚ T

6. 38,600 mg = ⬚ dag

7. 87 oz = ⬚ lb ⬚ oz

8. 0.65 T = ⬚ lb

Problem Solving

9. Maggie bought 52 ounces of swordfish selling for $6.92 per pound. What was the total cost?

10. Three bunches of grapes have masses of 1,000 centigrams, 1,000 decigrams, and 1,000 grams, respectively. What is the total combined mass of the grapes in kilograms?

11. **WRITE** ▶ *Math* Explain how you would find the number of ounces in 0.25T.

_____

_____

Lesson Check (6.RP.A.3d)

1. The mass of Denise's rock sample is 684 grams. The mass of Pauline's rock sample is 29,510 centigrams. How much greater is the mass of Denise's sample than Pauline's sample?

_____

2. A sign at the entrance to a bridge reads: Maximum allowable weight 2.25 tons. Jason's truck weighs 2,150 pounds. How much additional weight can he carry?

_____

Spiral Review (6.RP.A.1, 6.RP.A.2, 6.RP.A.3a, 6.RP.A.3b, 6.RP.A.3c)

3. There are 23 students in a math class. Twelve of them are boys. What is the ratio of girls to total number of students?

_____

4. Miguel hiked 3 miles in 54 minutes. At this rate, how long will it take him to hike 5 miles?

_____

5. Marco borrowed $150 from his brother. He has paid back 30% so far. How much money does Marco still owe his brother?

_____

6. How many milliliters are equivalent to 2.7 liters?

_____

FOR MORE PRACTICE
GO TO THE
Personal Math Trainer

Name _____

Vocabulary

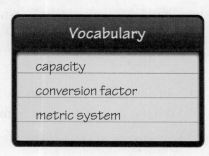

Choose the best term from the box to complete the sentence.

1. A _____ is a rate in which the two quantities are equal, but use different units. (p. 315)

2. _____ is the amount a container can hold. (p. 321)

Concepts and Skills

Convert units to solve. (6.RP.A.3d)

3. A professional football field is 160 feet wide. What is the width of the field in yards?

_____

4. Julia drinks 8 cups of water per day. How many quarts of water does she drink per day?

_____

5. The mass of Hinto's math book is 4,458 grams. What is the mass of 4 math books in kilograms?

_____

6. Turning off the water while brushing your teeth saves 379 centiliters of water. How many liters of water can you save if you turn off the water the next 3 times you brush your teeth?

_____

Convert to the given unit. (6.RP.A.3d)

7. 34.2 mm = _____ cm

8. 42 in. = _____ ft

9. 1.4 km = _____ hm

10. 4 gal = _____ qt

11. 53 dL = _____ daL

12. 28 c = _____ pt

13. Trenton's laptop is 32 centimeters wide. What is the width of the laptop in decimeters? (6.RP.A.3d)

_____

14. A truck is carrying 8 cars weighing an average of 4,500 pounds each. What is the total weight in tons of the cars on the truck? (6.RP.A.3d)

_____

15. GO DEEPER Ben's living room is a rectangle measuring 10 yards by 168 inches. By how many feet does the length of the room exceed the width? (6.RP.A.3d)

_____

16. Jessie served 13 pints of orange juice at her party. How many quarts of orange juice did she serve? (6.RP.A.3d)

_____

17. Kaylah's cell phone has a mass of 50,000 centigrams. What is the mass of her phone in grams? (6.RP.A.3d)

_____

Transform Units

Essential Question How can you transform units to solve problems?

You can sometimes use the units of the quantities in a problem to help you decide how to solve the problem.

Common Core Ratios and Proportional Relationships—6.RP.A.3d

MATHEMATICAL PRACTICES
MP1, MP3, MP6

Unlock the Problem

A car's gas mileage is the average distance the car can travel on 1 gallon of gas. Maria's car has a gas mileage of 20 miles per gallon. How many miles can Maria travel on 9 gallons of gas?

> • Would you expect the answer to be greater or less than 20 miles? Why?
>
> _____
>
> _____
>
> _____

Analyze the units in the problem.

STEP 1 Identify the units.

You know two quantities: the car's gas mileage and the amount of gas.

Gas mileage: 20 miles per gallon = $\dfrac{20\rule{2cm}{0.4pt}}{1}$

Amount of gas: 9 _____

You want to know a third quantity: the distance the car can travel.

Distance: ■ _____

_____

STEP 2 Determine the relationship among the units.

Think: The answer needs to have units of miles. If I multiply $\frac{20 \text{ miles}}{1 \text{ gallon}}$ by 9 gallons, I can simplify the units. The product will have units of

_____, which is what I want.

STEP 3 Use the relationship.

$$\frac{20 \text{ mi}}{1 \text{ gal}} \times 9 \text{ gal} = \frac{20 \text{ mi}}{1 \text{ \cancel{gal}}} \times \frac{9 \text{ \cancel{gal}}}{1} = \underline{\hspace{3cm}}$$

So, Maria can travel _____ on 9 gallons of gas.

1. Explain why the units of gallons are crossed out in the multiplication step above.

   _____

   _____

Sometimes you may need to convert units before solving a problem.

🔑 Example

The material for a rectangular awning has an area of 315 square feet. If the width of the material is 5 yards, what is the length of the material in feet? (Recall that the area of a rectangle is equal to its length times its width.)

STEP 1 Identify the units.

You know two quantities: the area of the material and the width of the material.

Area: 315 sq ft = 315 ft × ft

Width: 5 _____

You want to know a third quantity: the length of the material.

Length: ⬛ ft

> **Math Idea**
> You can write units of area as products.
> sq ft = ft × ft

STEP 2 Determine the relationship among the units.

Think: The answer needs to have units of feet. So, I should convert the width from yards to feet.

Width: $\dfrac{5 \text{ yd}}{1} \times \dfrac{\boxed{}\ \text{ft}}{1 \text{ yd}} = \boxed{}$ ft

Think: If I divide the area by the width, the units will simplify. The quotient will have units of _____, which is what I want.

STEP 3 Use the relationship.

Divide the area by the width to find the length.

315 sq ft ÷ _____ ft

Write the division using a fraction bar.

$\dfrac{\boxed{}\ \text{sq ft}}{15 \text{ ft}}$

Write the units of area as a product and divide the common units.

$\dfrac{\boxed{}\ \text{ft} \times \cancel{\text{ft}}}{\cancel{\text{ft}}} = \boxed{}$ ft

So, the length of the material is _____.

MATHEMATICAL PRACTICES ①
Analyze How can examining the units in a problem help you solve the problem?

2. **MATHEMATICAL PRACTICE ③** **Apply** Explain how knowing how to find the area of a rectangle could help you solve the problem above.

_____

_____

3. **MATHEMATICAL PRACTICE ⑥** **Explain** why the answer is in feet.

_____

_____

_____

Name _____

1. A dripping faucet leaks
12 gallons of water per day.
How many gallons does the faucet
leak in 6 days?

Quantities you know: $\dfrac{12}{1}$ _____ and _____ days

Quantity you want to know: ▨ _____

$\dfrac{\boxed{}\text{ gal}}{1\text{ day}} \times \boxed{}\text{ days} = $ _____

So, the faucet leaks _____ in 6 days.

2. Bananas sell for $0.44 per pound. How much will 7 pounds of bananas cost?

_____

3. Grizzly Park is a rectangular park with an area of 24 square miles. The park is 3 miles wide. What is its length in miles?

_____

On Your Own

Multiply or divide the quantities.

4. $\dfrac{24\text{ kg}}{1\text{ min}} \times 15\text{ min}$

_____

5. $216\text{ sq cm} \div 8\text{ cm}$

_____

6. $\dfrac{17\text{ L}}{1\text{ hr}} \times 9\text{ hr}$

_____

7. **GO DEEPER** The rectangular rug in Marcia's living room measures 12 feet by 108 inches. What is the rug's area in square feet?

_____

8. **MATHEMATICAL PRACTICE ①** Make Sense of Problems
A box-making machine makes cardboard boxes at a rate of 72 boxes per minute. How many minutes does it take to make 360 boxes?

_____

Personal Math Trainer

9. **THINK SMARTER +** The area of an Olympic-size swimming pool is 1,250 square meters. The length of the pool is 5,000 centimeters. Select True or False for each statement.

9a. The length of the pool is 50 meters. ○ True ○ False

9b. The width of the pool is 25 meters. ○ True ○ False

9c. The area of the pool is 1.25 square kilometers ○ True ○ False

Connect to Reading

Make Predictions

A *prediction* is a guess about something in the future. A prediction is more likely to be accurate if it is based on facts and logical reasoning.

The Hoover Dam is one of America's largest producers of hydroelectric power. Up to 300,000 gallons of water can move through the dam's generators every second. Predict the amount of water that moves through the generators in half of an hour.

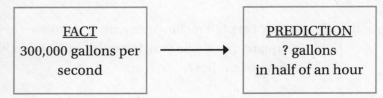

FACT	PREDICTION
300,000 gallons per second	? gallons in half of an hour

Use what you know about transforming units to make a prediction.

You know the rate of the water through the generators, and you are given an amount of time.

Rate of flow: $\dfrac{\quad \text{gal}}{1 \text{ sec}}$; time: $\dfrac{1}{2}$ _____

You want to find the amount of water.

Amount of water: ▪ gallons

Convert the amount of time to seconds to match the units in the rate.

$\dfrac{1}{2}$ hr = _____ min

$\dfrac{30 \text{ min}}{1} \times \dfrac{\quad \text{sec}}{1 \text{ min}} = \underline{\hspace{2cm}}$ sec

Multiply the rate by the amount of time to find the amount of water.

$\dfrac{\quad \text{gal}}{\text{sec}} \times \dfrac{\quad \cdot \text{ sec}}{1} = \underline{\hspace{2cm}}$ gal

So, a good prediction of the amount of water that moves through the generators in half of an hour is _____.

Transform units to solve.

10. An average of 19,230 people tour the Hoover Dam each week. Predict the number of people touring the dam in a year.

_____

11. **THINK SMARTER** The Hoover Dam generates an average of about 11,506,000 kilowatt-hours of electricity per day. Predict the number of kilowatt-hours generated in 7 weeks.

_____

Name _____

Transform Units

Common Core

COMMON CORE STANDARD—6.RP.A.3d
Understand ratio concepts and use ratio reasoning to solve problems.

Multiply or divide the quantities.

1. $\dfrac{62\,g}{1\,day} \times 4$ days

   $\dfrac{62\,g}{1\,\cancel{day}} \times \dfrac{4\,\cancel{days}}{1} = 248$ g

2. 322 sq yd ÷ 23 yd

   $\dfrac{322\,sq\,yd}{23\,yd}$

   $\dfrac{322\,yd \times \cancel{yd}}{23\,\cancel{yd}} = 14$ yd

3. $\dfrac{128\,kg}{1\,hr} \times 10$ hr

4. 136 sq km ÷ 8 km

5. $\dfrac{88\,lb}{1\,day} \times 12$ days

6. 154 sq mm ÷ 11 mm

7. $\dfrac{\$150}{1\,sq\,ft} \times 20$ sq ft

8. 234 sq ft ÷ 18 ft

Problem Solving Real World

9. Green grapes are on sale for $2.50 a pound. How much will 9 pounds cost?

10. A car travels 32 miles for each gallon of gas. How many gallons of gas does it need to travel 192 miles?

11. **WRITE** ▸ *Math* Write and solve a problem in which you have to transform units. Use the rate 45 people per hour in your question.

_____

_____

_____

Lesson Check (6.RP.A.3d)

1. A rectangular parking lot has an area of 682 square yards. The lot is 22 yards wide. What is the length of the parking lot?

2. A machine assembles 44 key chains per hour. How many key chains does the machine assemble in 11 hours?

Spiral Review (6.RP.A.3a, 6.RP.A.3c)

3. Three of these ratios are equivalent to $\frac{8}{20}$. Which one is NOT equivalent?

$$\frac{2}{5} \qquad \frac{12}{24} \qquad \frac{16}{40} \qquad \frac{40}{100}$$

4. The graph shows the money that Marco earns for different numbers of days worked. How much money does he earn per day?

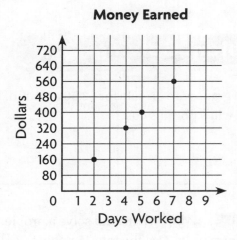

5. Megan answered 18 questions correctly on a test. That is 75% of the total number of questions. How many questions were on the test?

FOR MORE PRACTICE
GO TO THE
Personal Math Trainer

Name _____

Problem Solving • Distance, Rate, and Time Formulas

Essential Question How can you use the strategy *use a formula* to solve problems involving distance, rate, and time?

Common Core
Ratios and Proportional Relationships—6.RP.A.3d
MATHEMATICAL PRACTICES
MP1, MP3, MP7

You can solve problems involving distance, rate, and time by using the formulas below. In each formula, *d* represents distance, *r* represents rate, and *t* represents time.

Distance, Rate, and Time Formulas		
To find distance, use $d = r \times t$	To find rate, use $r = d \div t$	To find time, use $t = d \div r$

🗝 Unlock the Problem

Helena drives 220 miles to visit Niagara Falls. She drives at an average speed of 55 miles per hour. How long does the trip take?

Use the graphic organizer to help you solve the problem.

Read the Problem

What do I need to find?

I need to find the _____ the trip takes.

What information do I need to use?

I need to use the _____ Helena travels and

the _____ of speed her car is moving.

How will I use the information?

First I will choose the formula _____ because I

need to find time. Next I will substitute for *d* and *r*. Then

I will _____ to find the time.

Solve the Problem

- First write the formula for finding time.

  $t = d \div r$

- Next substitute the values for *d* and *r*.

  $t =$ _____ mi $\div \dfrac{\boxed{}\ mi}{1\ hr}$

- Rewrite the division as multiplication by the reciprocal of $\dfrac{55\ mi}{1\ hr}$.

  $t = \dfrac{\boxed{}\ \cancel{mi}}{1} \times \dfrac{1\ hr}{\boxed{}\ \cancel{mi}} =$ _____ hr

MATHEMATICAL PRACTICES ⑦

Look for Structure How do you know which formula to use?

So, the trip takes _____ hours.

🕐 Try Another Problem

Santiago's class traveled to the Museum of Natural Science for a field trip. To reach the destination, the bus traveled at a rate of 65 miles per hour for 2 hours. What distance did Santiago's class travel?

Choose a formula.

$$d = r \times t \qquad r = d \div t \qquad t = d \div r$$

Use the graphic organizer below to help you solve the problem.

Read the Problem	Solve the Problem
What do I need to find?	
What information do I need to use?	
How will I use the information?	

So, Santiago's class traveled _____ miles.

Math Talk

MATHEMATICAL PRACTICES ①

Evaluate How could you check your answer by solving the problem a different way?

1. **What if** the bus traveled at a rate of 55 miles per hour for 2.5 hours? How would the distance be affected?

_____

2. **MATHEMATICAL PRACTICE ⑦ Identify Relationships** Describe how to find the rate if you are given the distance and time.

_____

_____

Name _____

Unlock the Problem

√ Choose the appropriate formula.
√ Include the unit in your answer.

1. Mariana runs at a rate of 180 meters per minute. How far does she run in 5 minutes?

   First, choose a formula.

   _____

   Next, substitute the values into the formula and solve.

   _____

   So, Mariana runs _____ in 5 minutes.

2. **THINK SMARTER** **What if** Mariana runs for 20 minutes at the same speed? How many kilometers will she run?

   _____

3. A car traveled 130 miles in 2 hours. How fast did the car travel?

   _____

4. A subway car travels at a rate of 32 feet per second. How far does it travel in 16 seconds?

   _____

5. A garden snail travels at a rate of 2.6 feet per minute. At this rate, how long will it take for the snail to travel 65 feet?

   _____

6. **GO DEEPER** A squirrel can run at a maximum speed of 12 miles per hour. At this rate, how many seconds will it take the squirrel to run 3 miles?

   _____

7. **THINK SMARTER** A cyclist rides 8 miles in 32 minutes. What is the speed of the cyclist in miles per hour?

   _____

WRITE ▸*Math* · **Show Your Work**

© Houghton Mifflin Harcourt Publishing Company

On Your Own

8. A pilot flies 441 kilometers in 31.5 minutes. What is the speed of the airplane?

_____

9. **GO DEEPER** Chris spent half of his money on a pair of headphones. Then he spent half of his remaining money on CDs. Finally, he spent his remaining $12.75 on a book. How much money did Chris have to begin with?

_____

WRITE ▸ Math
Show Your Work

10. **THINK SMARTER** André and Yazmeen leave at the same time and travel 75 miles to a fair. André drives 11 miles in 12 minutes. Yazmeen drives 26 miles in 24 minutes. If they continue at the same rates, who will arrive at the fair first? Explain.

_____

11. **MATHEMATICAL PRACTICE ③ Make Arguments** Bonnie says that if she drives at an average rate of 40 miles per hour, it will take her about 2 hours to drive 20 miles across town. Does Bonnie's statement make sense? Explain.

_____

Personal Math Trainer

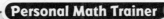

12. **THINK SMARTER +** Claire says that if she runs at an average rate of 6 miles per hour, it will take her about 2 hours to run 18 miles. Do you agree or disagree with Claire? Use numbers and words to support your answer.

_____

_____

_____

_____

Name _____

Problem Solving • Distance, Rate, and Time Formulas

COMMON CORE STANDARD—6.RP.A.3d
Understand ratio concepts and use ratio reasoning to solve problems.

Read each problem and solve.

1. A downhill skier is traveling at a rate of 0.5 mile per minute. How far will the skier travel in 18 minutes?

   $d = r \times t$

   $d = \dfrac{0.5 \text{ mi}}{1 \text{ min}} \times 18 \text{ min}$

   $d = 9 \text{ miles}$

   _____

2. How long will it take a seal swimming at a speed of 8 miles per hour to travel 52 miles?

   _____

3. A dragonfly traveled at a rate of 35 miles per hour for 2.5 hours. What distance did the dragonfly travel?

   _____

4. A race car travels 1,212 kilometers in 4 hours. What is the car's rate of speed?

   _____

5. Kim and Jay leave at the same time to travel 25 miles to the beach. Kim drives 9 miles in 12 minutes. Jay drives 10 miles in 15 minutes. If they both continue at the same rate, who will arrive at the beach first?

   _____

6. **WRITE** ▸*Math* Describe the location of the variable *d* in the formulas involving rate, time, and distance.

   _____

   _____

   _____

Lesson Check (6.RP.A.3d)

1. Mark cycled 25 miles at a rate of 10 miles per hour. How long did it take Mark to cycle 25 miles?

2. Joy ran 13 miles in $3\frac{1}{4}$ hours. What was her average rate?

Spiral Review (6.RP.A.3a, 6.RP.A.3c, 6.RP.A.3d)

3. Write two ratios that are equivalent to $\frac{9}{12}$.

4. In the Chang family's budget, 0.6% of the expenses are for internet service. What fraction of the family's expenses is for internet service? Write the fraction in simplest form.

5. How many meters are equivalent to 357 centimeters?

6. What is the product of the two quantities shown below?

$$\frac{60 \text{ mi}}{1 \text{ hr}} \times 12 \text{ hr}$$

FOR MORE PRACTICE
GO TO THE
Personal Math Trainer

✓ Chapter 6 Review/Test

Personal Math Trainer
Online Assessment
and Intervention

1. A construction crew needs to remove 2.5 tons of river rock during the construction of new office buildings.

   The weight of the rocks is

800
2,000
5,000

   pounds.

2. Select the conversions that are equivalent to 10 yards.
   Mark all that apply.

   (A) 20 feet   (C) 30 feet

   (B) 240 inches   (D) 360 inches

3. Meredith runs at a rate of 190 meters per minute. Use the formula $d = r \times t$ to find how far she runs in 6 minutes.

4. The table shows data for 4 cyclists during one day of training. Complete the table by finding the speed for each cyclist. Use the formula $r = d \div t$.

Cyclist	Distance (mi)	Time (hr)	Rate (mi per hr)
Alisha	36	3	
Jose	39	3	
Raul	40	4	
Ruthie	22	2	

5. For numbers 5a–5c, choose <, >, or =.

5a. 5 kilometers [< > =] 5,000 meters

5b. 254 centiliters [< > =] 25.4 liters

5c. 6 kilogram [< > =] 600 gram

6. A recipe calls for 16 fluid ounces of light whipping cream. If Anthony has 1 pint of whipping cream in his refrigerator, does he have enough for the recipe? Explain your answer using numbers and words.

7. For numbers 7a–7d, choose <, >, or =.

7a. 43 feet [< > =] 15 yards   7c. 10 pints [< > =] 5 quarts

7b. 5 tons [< > =] 5000 pounds   7d. 6 miles [< > =] 600 yards

8. **GO DEEPER** The distance from Caleb's house to the school is 1.5 miles, and the distance from Ashlee's house to the school is 3,520 feet. Who lives closer to the school, Caleb or Ashlee? Use numbers and words to support your answer.

348

9. Write the mass measurements in order from least to greatest.

7.4 kilograms	7.4 decigrams	7.4 centigrams

_____   _____   _____

10. An elephant's heart beats 28 times per minute. Complete the product to find how many times its heart beats in 30 minutes.

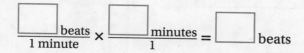

$$\frac{\boxed{}\text{ beats}}{1 \text{ minute}} \times \frac{\boxed{}\text{ minutes}}{1} = \boxed{}\text{ beats}$$

11. The length of a rectangular football field, including both end zones, is 120 yards. The area of the field is 57,600 square feet. For numbers 11a–11d, select True or False for each statement.

11a. The width of the field is 480 yards.   ○ True   ○ False

11b. The length of the field is 360 feet.   ○ True   ○ False

11c. The width of the field is 160 feet.   ○ True   ○ False

11d. The area of the field is 6,400 square yards.   ○ True   ○ False

12. Harry received a package for his birthday. The package weighed 357,000 centigrams. Select the conversions that are equivalent to 357,000 centigrams. Mark all that apply.

○ 3.57 kilograms

○ 357 dekagrams

○ 3,570 grams

○ 3,570,000 decigrams

13. Mr. Martin wrote the following problem on the board.

> Juanita's car has a gas mileage of 21 miles per gallon. How many miles can Juanita travel on 7 gallons of gas?

Alex used the expression $\frac{21 \text{ miles}}{1 \text{ gallon}} \times \frac{1}{7 \text{ gallons}}$ to find the answer. Explain

Alex's mistake.

14. Mr. Chen filled his son's wading pool with 20 gallons of water.

20 gallons is equivalent to $\begin{array}{|c|} \hline 80 \\ \hline 60 \\ \hline 40 \\ \hline \end{array}$ quarts.

15. Nadia has a can of vegetables with a mass of 411 grams. Write equivalent conversions in the correct boxes.

| 4.11 | | 41.1 | | 0.411 |

kilograms	hectograms	dekagrams

16. Steve is driving 440 miles to visit the Grand Canyon. He drives at an average rate of 55 miles per hour. Explain how you can find the amount of time it will take Steve to get to the Grand Canyon.

17. Lucy walks one time around the lake. She walks for 1.5 hours at an average rate of 3 miles per hour. What is the distance, in miles, around the lake?

_____ miles

18. The parking lot at a store has a width of 20 yards 2 feet and a length of 30 yards.

20 yards 2 feet

30 yards

Part A

Derrick says that the width could also be written as 22 feet.
Explain whether you agree or disagree with Derrick.

Part B

The cost to repave the parking lot is $2 per square foot. Explain how much it would cost to repave the parking lot.

19. THINK SMARTER + Jake is using a horse trailer to take his horses to his new ranch.

    Part A

    Complete the table by finding the weight, in pounds, of Jake's horse trailer and each horse.

	Weight (T)	Weight (lb)
Horse	0.5	
Trailer	1.25	

    Part B

    Jake's truck can tow a maximum weight of 5,000 pounds. What is the maximum number of horses he can take in his trailer at one time without going over the maximum weight his truck can tow? Use numbers and words to support your answer.

20. A rectangular room measures 13 feet by 132 inches. Tonya said the area of the room is 1,716 square feet. Explain her mistake, then find the area in square feet.